Mathematics Encyclopedia

Everyone uses some form of mathematics every day. When they make purchases, tell the time, catch a bus or measure the floor for a new carpet.

Questions like
 How many?
  How much?
   How big?
    How often?
     How far?
      How fast?

are answered by knowing how numbers relate to each other and how different parts of space fit together.

Mathematics is a way of thinking. It helps people to solve problems and understand the world in which they live. It has its own language and vocabulary, some of which this book will help you to understand. They are keys to an exciting store of knowledge.

Mathematics Encyclopedia

Patricia and Victor Smeltzer

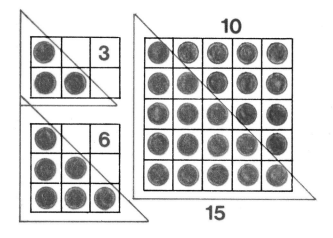

Edited by Audrey Sutton
Illustrated by Sue Sharples

Burke Books LONDON * TORONTO * NEW YORK

First published July 1980
First published in Canada August 1980
First published in the United States of America August 1980
© Patricia and Victor Smeltzer 1980
Illustrations © Burke Publishing Company Limited 1980

All rights reserved. No part of this publication may be
reproduced, stored in a retrieval system, or transmitted
in any form or by any means, electronic, mechanical,
photocopying, recording or otherwise, without the prior
permission of Burke Publishing Company Limited, Burke
Publishing (Canada) Limited or Burke Publishing Company Inc.

Library of Congress Catalog Card Number 80/40691

New edition May 1983

Smeltzer, Patricia
  Mathematics encyclopedia.
  1. Mathematics – Dictionaries, Juvenile
  I. Title II. Smeltzer, Victor III. Sutton,
Audrey
510'.3 QA5
ISBN 0 222 00692 7 Library
ISBN 0 222 00694 3 Softback

Burke Publishing Company Limited
Pegasus House, 116–120 Golden Lane, London EC1Y 0TL, England.
Burke Publishing (Canada) Limited
91 Station Street, Ajax, Ontario L1S 3H2, Canada.
Burke Publishing Company Inc.,
1 Emerald Street, Norwalk, Conn. 06850, U.S.A.
Printed in Italy by Vallardi Industrie Grafiche, Lainate, (Milan).

Contents

Encyclopedia	6
abacus–axis	6
balance–bushel	11
calculus–cylinder	13
day–dodecahedron	20
edge–even number	24
face–furlong	27
gallon–gross	29
hand–hypotenuse	33
icosahedron–isosceles triangle	36
Kelvin scale–knot	38
latitude–longitude	39
"magic" square–multiplication	41
natural number–numerator	46
oblique–ounce	48
palm–Pythagorean theorem	50
quadrant–quotient	59
radius–rood	61
scale–symmetry	65
tally–two-dimensional	74
undecagon–universal set	80
variable–volume	81
watt–width	82
yard–year	84
zero	84
The Story of Mathematics	85
SI, Metric and Imperial Tables	89
Index	93

abacus

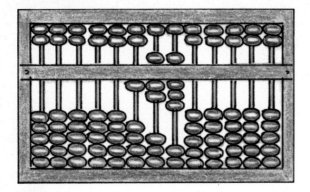

A counting frame with rods, on which calculations are made by moving sliding balls or beads. It was used by the ancient Egyptians, Chinese, Greeks and Romans. It is used today by the Chinese and also for teaching the first rules of counting to children in countries throughout the world.

acre

In olden times an acre was the amount of land that could be ploughed by one man in one day. The measure was standardised, by King Henry VIII, at 4,840 square yards.

$5 + 5 = 10$

$7 + 2 = 9$

add/addition

Increase by putting units together. The mathematical sign is + (plus).

algebra

A branch of mathematics in which symbols, usually letters, stand for numbers. It is a code for writing mathematical problems in the shortest, clearest way so that one letter can tell you something about several numbers or even all numbers. This method was developed by the Arabs.

Let 'a' stand for any number and 'b' for any other number

$a = 2 \quad b = 4 \quad a + b = b + a$

algebraic equation

In algebra problems are usually made into the form of equations. An equation resembles a pair of scales in perfect balance. Both sides of the scales must be changed in the same way for it to balance.

altitude

1) Height above sea level.

2) The perpendicular distance from the base of a triangle to its topmost point.

angle

The amount of rotation (turning) of a straight line round a point from one direction to another. The unit used to measure the angle is called a degree (°). The symbol for an angle is ∟.

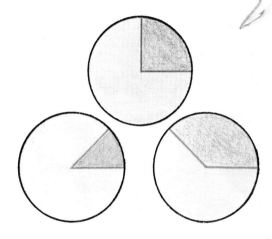

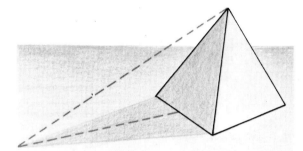

angle of elevation

The angle between the horizon and an imaginary line joining the horizon at a particular point to the topmost point of an object.

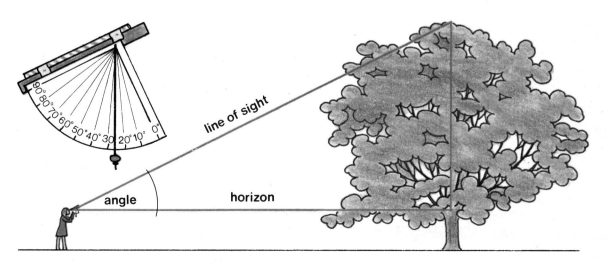

angle-meter

A semi-circular dial – marked up to 180° – used in drawing offices to fix the position of one object or point in relation to another by measuring the angle.

ante meridiem

Literally, before midday (in Latin). When the 24-hour clock is not used, the abbreviation a.m. is used to show a time in the morning, between midnight and midday.

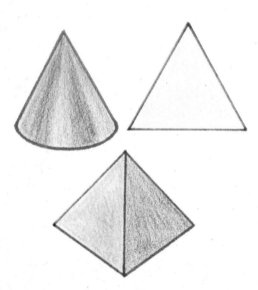

apex

The top, pointed tip or peak.

approximation

A way of coming close to an accurate number, used when there is not enough information for a completely accurate number to be reached, or instead of long calculations when an exact answer is not needed.

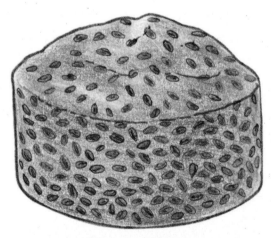

guess how many currants in the cake?

Arabic numerals

Ten symbols developed by the Arabs and with which any number can be written down. The Arabs took the idea of having a symbol for zero from the ancient people of India.

1 2 3 4 5
6 7 8 9 0

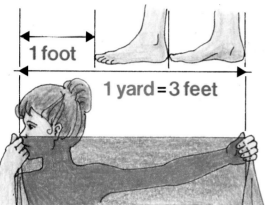

arbitrary measures

Measures that are not bound by rules and therefore will not always mean the same; e.g. this ball balances eight marbles (if different marbles were used you might need more or less than eight); or, the room is ten paces wide.

arc

A part of a circle between any two points on that circle's circumference.

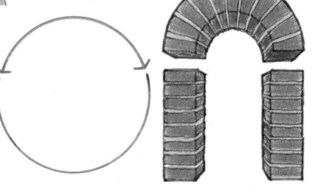

area

A surface or region inside a given boundary. It is two-dimensional and measured in square units. A quick way of measuring the area of a square or rectangle is to use the formula: Area=length×width.

The SI symbol is A and the SI unit of area is the metre square (m²).

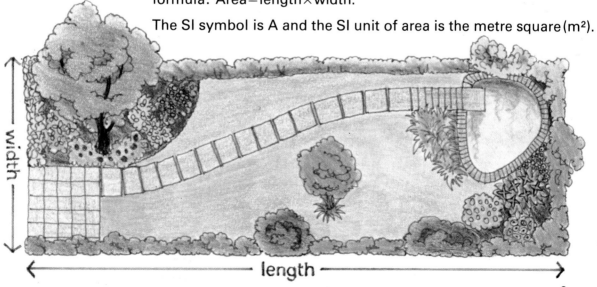

arithmetic

The study of numbers and counting. The four basic operations of arithmetic are addition, subtraction, multiplication and division.

$5+7=12 \quad 9-4=5$
$8\times2=16 \quad 18\div3=6$

astronomy

The science of the stars – in particular the study of their positions and movements. People have been interested in the "heavenly bodies" since as early as the seventh century B.C. but the pioneers of modern astronomy were Copernicus (in the sixteenth century), Galileo and Newton. Today astronomy has developed much further to include the study of systems far outside our own galaxy.

asymmetry

Lack of balance or proporation. An object or figure that cannot be divided into two exactly equal or opposite parts in asymmetrical.

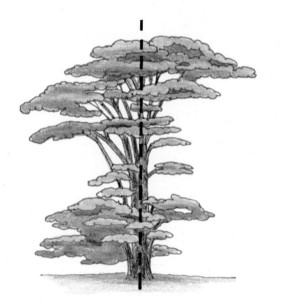

atomic clock

The most accurate means of measuring time. The atomic clock is based on the very rapid rhythms inside an atom. With this clock irregularities in the spinning of the earth can be measured. There is one atomic clock in England, at the National Physical Laboratory in Middlesex.

average

A value obtained by adding together all the quantities in a set and dividing the total by their number. Often used as a rough standard; e.g. the average height of children of a certain age.

This child is of average height

axis

1) The line dividing a symmetrical shape into equal parts — as in axis of symmetry.

2) An imaginary line on which a solid figure rotates.

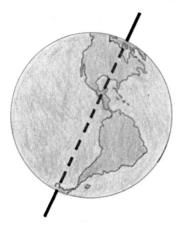

balance

A weighing instrument, usually consisting of a central beam with a pan on either arm. Used to weigh objects against each other or against standard weights.

bar graph

A method of showing statistics by the length of columns (bars).

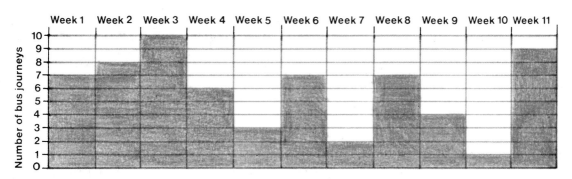

base

1) The side or face on which objects stand; the line forming the bottom part of a figure.

2) The basic number linking a system of counting. It tells how many units of one kind are required to form a new group. The most common way of counting in everyday life uses the base of 10; figures in each column being worth ten times those in the column to their right. To count in fives is Base Five, to count in threes is Base Three, etc.

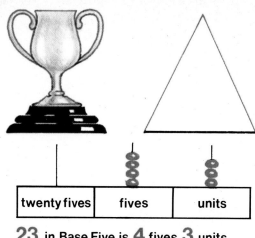

23 in Base Five is 4 fives 3 units
i.e. 43 five

bearing

Direction; position of one object relative to another.

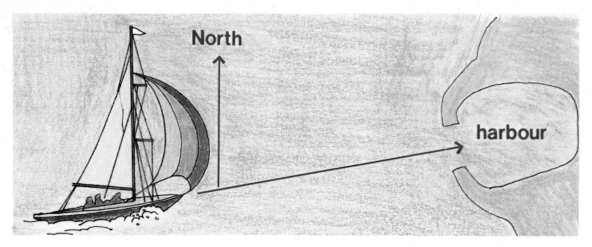

billion

A very large number which has a different value in Europe from that in the U.S.A. In Europe it means a million × million. In the U.S.A. it means a thousand × million.

binary system

A system of counting to the base of two. It is important because of its use in electronic computers which use a series of electric circuits. They build up large numbers by turning current on and off in these circuits.

Every number can be shown as 1 or 0

1 = 1   6 = 110
2 = 10   7 = 111
3 = 11   8 = 1000
4 = 100   9 = 1001
5 = 101   10 = 1010

bisect

Divide into two (usually equal) parts. A line that bisects is called a bisector.

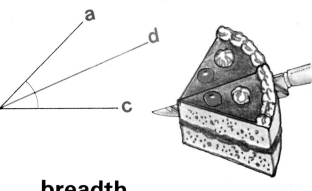

breadth

Width; the distance from one side to the other of a shape.

bushel

An old measure used for dry goods such as grain or fruit.

calculus

A branch of advanced mathematics that deals with things that move or change and their speed or rate of change.

calendar

A system by which the year is fixed and divided into months, weeks and days. There are two types — the solar calendar, which is divided according to the position of the sun at the different seasons, and the lunar calendar which is divided according to the phases of the moon. The lunar calendar is used by the Jews and many Eastern people.

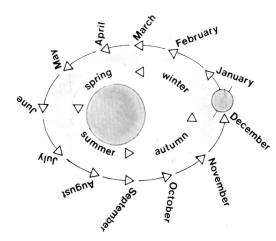

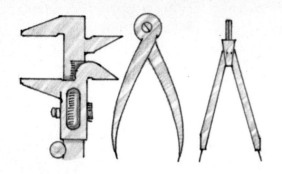

calliper

An instrument which looks rather like compasses, used for measuring the inside or outside diameter of objects, also to measure thickness and width.

candela

SI base unit of the intensity of light. The name comes from measuring the light of lamps by the number of candles they replaced. The symbol is cd.

capacity

The amount that can be held in a solid container. The SI unit of capacity for liquid measurement is the litre (l).
1 litre=1000 cubic centimetres (SI symbol cm³).
1 litre=1000 millilitres (ml).

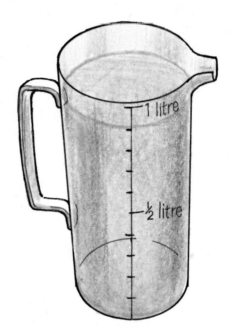

carat

See troy weight.

cardinal number

Any number answering the question "how many?" e.g. 1, 2, 3.

cardinal points

The four chief points of the compass: North, South, East and West (N, S, E and W).

centigrade or Celsius scale

The scale of temperature, as measured by a centigrade thermometer (first made by Celsius 1701–44), on which the freezing-point of pure water is 0° and the boiling-point is 100°. This scale uses the SI symbol °C.
(To convert to Fahrenheit, multiply the centigrade temperature by 9, divide the product by 5 and add 32.)

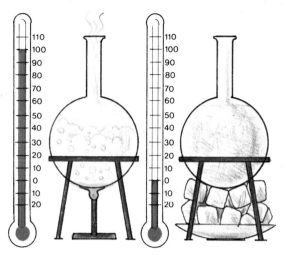

centilitre

A unit of liquid measurement equal to one hundredth part of a litre.
SI symbol cl.

centimetre

A unit of length equal to one hundredth of a metre. SI symbol cm.

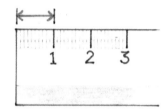

centre

The middle point.

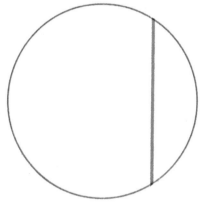

chord

A straight line joining any two points on the circumference of a circle or closed curve.

chronometer

A clock that keeps very accurate time over a long period and is not affected by changes in temperature. It was invented by John Harrison, an English carpenter, and is used by sailors for working out longitude. Using a sextant, the navigator can find local time accurately by measuring the amount the earth has turned towards the sun. Using a chronometer, he can carry Greenwich mean-time with him. The difference between the two times will show how far west or east he is of the Greenwich meridian.

circle

A closed plane (flat) curve, all of whose points are the same distance from a given centre point.

circumference

The set of points which make up the curved boundary line of a circle.

clinometer

Any of the various instruments used for finding heights by measuring angles in a vertical plane. Can be a semi-circular dial marked up to 180°. A simple clinometer can be made from a protractor and a plumb-line. A right-angled triangle made of cardboard or wood can also be used for measuring height. Clinometers are used in ships to show how much the ship is listing to one side or the other.

closed plane figure

A flat surface enclosed by one of the following:
1) one curve

2) two or more curve segments

3) three or more straight lines

4) one (or more than one) curve segment together with one (or more than one) straight line.
The points inside the figure are separated from the points outside it in the same plane by the lines.

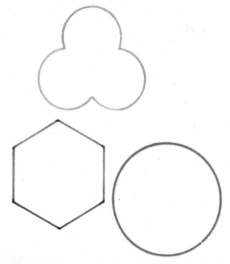

closed solid figure

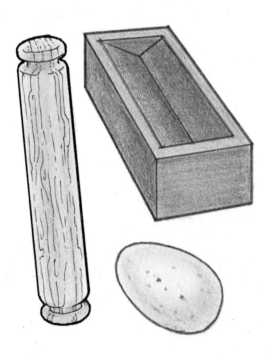

A solid figure so made that the points inside the solid are separated from the points outside. A football is a closed solid figure; a milk bottle without a top is an open solid figure.

common property

A quality which belongs to two or more objects; e.g. a tall boy and a tall girl have the common property of being tall; a red book and a blue book have the common property of being books.

compass

1) An instrument made with two hinged legs, one of which has a pencil attached, for drawing circles. A crude substitute can be made with a pin and a length of string.

2) A magnetic compass is an instrument consisting of a magnetised needle which always points to the magnetic north pole and a dial marked with all the points. By moving the compass round until the needle is exactly over the direction marked as magnetic north on the dial, other directions can be found.

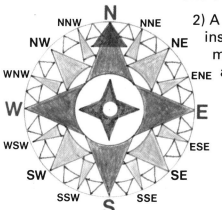

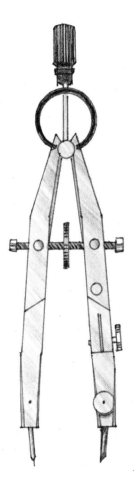

composite number

A number which has factors; i.e. can be obtained by multiplying two whole numbers other than itself and one. Any natural number that is not a prime number must be a composite number.

e.g. 6 has the factors 2, 3
The series begins
4, 6, 8, 10, 12, 14,
15, 16, 18, 20 etc.

concave

Curving inwards, like a cave (the opposite of convex).

cone

A solid formed by turning a right-angled triangle around one of the sides which has the right angle. Sections through cones give three important curves, the ellipse, the parabola and the hyperbola.

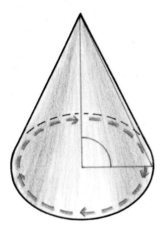

conservation

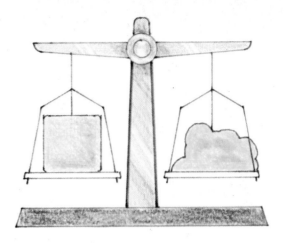

The property of remaining the same while appearing different; e.g. 500 g of butter which has been squashed has the same weight as it had before it was squashed – the weight is conserved.

continuous graph

A graph in which the known points can all be joined by a straight line. This means that the whole line can be used as a ready reckoner and the corresponding values read off for any point on it.

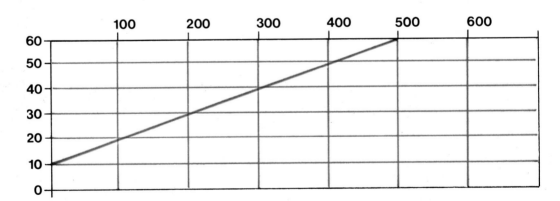

convex

Curving outwards (the opposite of concave).

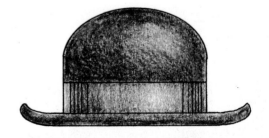

co-ordinates

The two numbers which indicate the exact position of a point on a map or graph. It is usual to take the first number from the set of numbers across the page and the second number from the set going up the page.

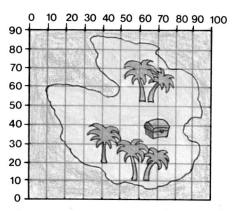

position of treasure (70,40)

count

To name the numerals in correct order so as to find out how many there are of a certain set.

cube

A six-sided closed solid in which all the faces (sides) are squares equal in all respects. A square box shape.

cubit

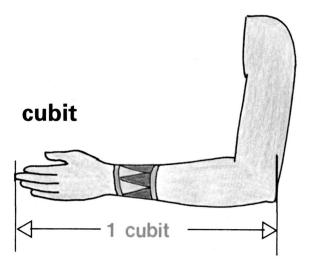

A unit of measure used in the ancient world and frequently mentioned in the Bible. The distance from the tip of the elbow to the tip of the middle finger (45–55 cm).

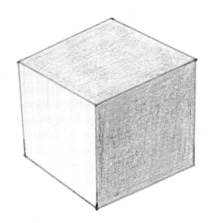

cuboid

A rectangular box shape. A six-sided closed solid whose opposite faces (sides) are parallel and equal in all respects.

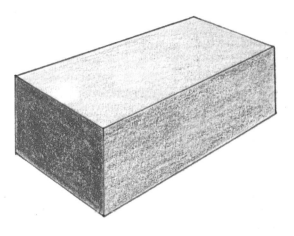

cylinder

A roller-like solid, of equal diameter throughout its length, in which the two end faces are circles which are parallel and equal in all respects.

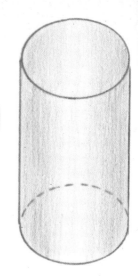

day

A unit of time, twenty-four hours. It is the time the earth takes to revolve once on its own axis. The SI symbol is d.

dead reckoning

The oldest system of navigation. A very rough method of finding position by working out how far the ship has gone since the last known position (often the home port). This is done by checking the ship's speed and the direction of the compass regularly, and marking the rough course on a chart.

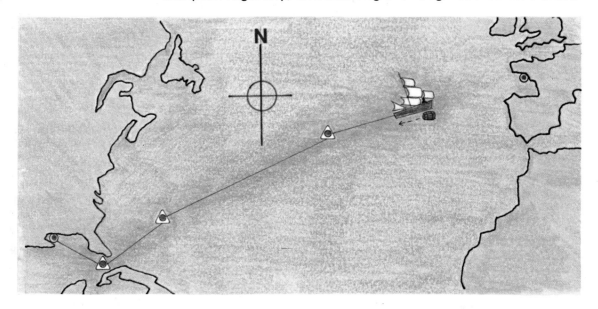

decade

A period of ten years.

1980 1990 2000 2010 2020 2030

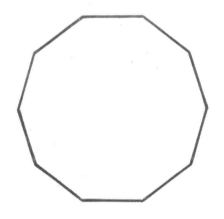

decagon

A plane (flat) figure with ten sides and ten angles. A regular decagon has ten equal sides and ten equal angles.

decilitre

A measure of volume, equal to one tenth of a litre. The SI symbol is dl.

decimal notation

Numbering in sets of ten. The decimal point is used to extend the whole-number column. Each column to the left is ten times the value of the numbers to the right.

$$\text{HTU} \cdot \tfrac{1}{10} \ \tfrac{1}{100} \ \tfrac{1}{1000}$$
$$3 \cdot 3 = 3\tfrac{3}{10}$$
$$3 \cdot 03 = 3\tfrac{3}{100}$$

decimetre

A measure of distance, equal to one tenth of a metre. The SI symbol is dm.

degree

1) A unit for measuring angles. Any of 360 equally spaced lines radiating from the centre of a circle in the same plane is at an angle of one degree (1°) to the lines next to it.

2) A unit for measuring temperature. The usual unit of temperature is now the degree Celsius/centigrade (SI symbol °C) but the Fahrenheit scale is still used sometimes (SI symbol °F).

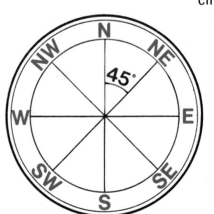

denominator

The number written under the line in a fraction. It names the number of parts into which a whole number has been divided. When two or more fractions are added or subtracted, a "common denominator" must be found – a multiple of each denominator concerned; e.g.
$$\tfrac{1}{2}+\tfrac{1}{3}=\tfrac{3+2}{6}=\tfrac{5}{6}$$

The top number of a fraction is called the numerator.

diagonal

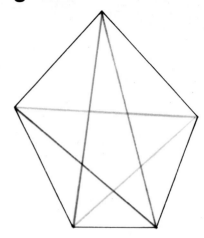

A line joining two angles of a rectangle, or other four- or more-sided figure, but which does not form a side of the shape.

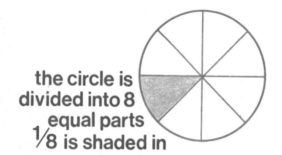

the circle is divided into 8 equal parts
$\tfrac{1}{8}$ is shaded in

diagram

A way of presenting information pictorially, as a plan or drawing in outline, to make understanding easier. There are various types of diagram; e.g. Venn.

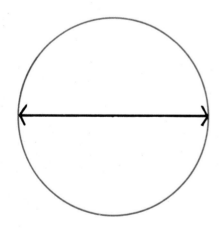

diameter

1) A straight line which passes through the centre of a circle dividing it into two equal parts.

2) The length of a straight line passing through the centre of a circular object from one side to the other.

digit

Any one of the ten symbols used to write numbers.

0 5 18 7 20 152
24 3 46 91 38
1 16 34 27 9 8

digital machine

Any machine which shows up information in digits; e.g. digital watches which use figures instead of the conventional clock face; desk calculators and counting machines which make all their calculations by repeated simple counting.

dimension

Any of the three measures of length, breadth and height. A line has one dimension (length); a plane figure has two dimensions (length and breadth); a solid figure has three dimensions (length, breadth and height).

disc

When the circumference and all the points within the circle are considered, the whole set is called a disc.

division

The process of finding how many times one number is contained in another. There are two ways of looking at division:
1) as repeated subtraction (this is the only way some computers can divide)
2) as separating objects into sets of equal size.
Division is the opposite of multiplication. For every division there is a multiplication that cancels the division. The sign is ÷.

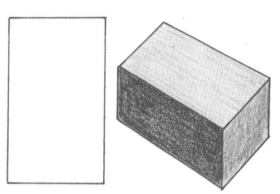

divisor   The number by which another is divided.

dodecagon

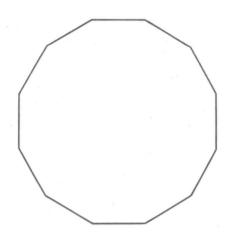

A plane figure with twelve sides and angles. A regular dodecagon has twelve equal sides and angles.

dodecahedron

A solid with twelve plane faces. A regular dodecahedron has twelve plane faces, each in the shape of a pentagon.

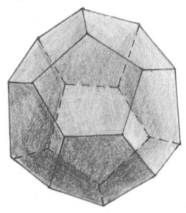

edge

The meeting line of two surfaces of a solid shape.

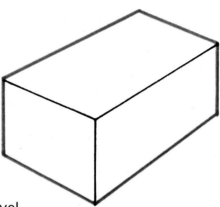

elevation   1) Height above a given level.

2) The angle made with the horizon by, for example, a jet of water from a hose.

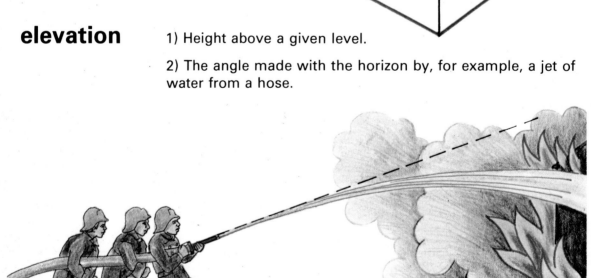

ellipse

An oval figure. You can make an ellipse by tying a loop of string loosely around two fixed drawing-pins, placing a pencil against the string on the inside so as to pull the loop tight and then drawing round the pins, always keeping the string tight. The orbit of the moon round the earth, and of any man-made satellite sent from earth, follows the path of an ellipse.

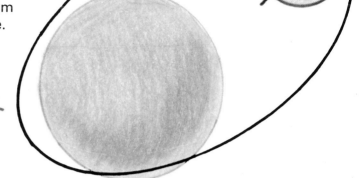

empty set

A set without any members. Symbol ∅ or { }.

enumerate

To count.

equation

A statement or sentence in mathematics in which the equality symbol (=) is used; e.g. 5+5=10. When a letter is used for an unknown number, the sentence becomes a problem; e.g. 5+a=10. When the number which replaces the unknown "a" is worked out, the problem is solved.

$$5+5=10 \quad 5+a=10$$

equiangular

Having equal angles.

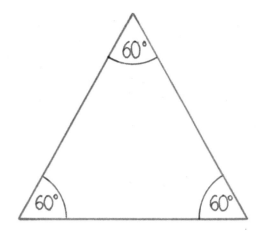

equidistant

At an equal distance from a specified point.

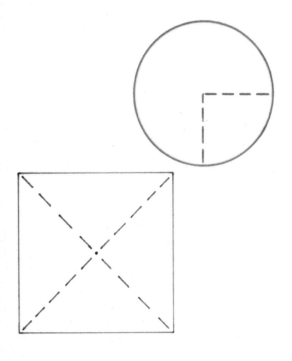

equilateral

Having all sides of equal length.

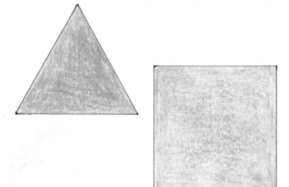

equipoise

Even balance achieved by making weights equal to each other.

equivalent

Equal in value.

$3/4 = 6/8$ $4 \times 250 \text{ ml.} = 1 \text{ l.}$

estimate

A calculated guess, or approximate opinion of an amount, made when there is not enough information for an accurate result or when an exact answer is not needed.

Guess how many spots on the dress?

0 2 4 6 8 10 12
30 56 70 98 102

even number

The number two or any positive whole number which is a multiple of two; a number which is divisible by two to give a whole number result.

face

One of the surfaces which make up a solid shape. A cube has six square faces.

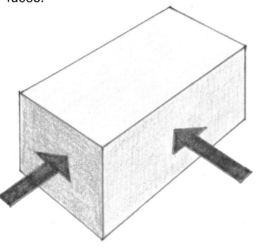

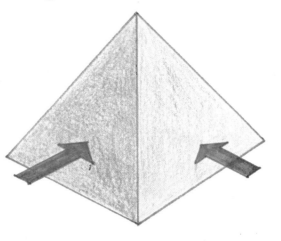

factor

One of the numbers which will divide equally into the given number; e.g. 4 and 3 are factors of 12.

Fahrenheit scale

The scale of temperature used with the imperial system. The freezing-point of pure water is 32°F, boiling-point is 212°F.
(To convert Fahrenheit to centigrade/Celsius, subtract 32, multiply by 5 and divide the product by 9.)

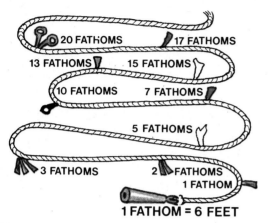

fathom

A unit of linear measurement (equal to 6 ft.) used to record the depth of water.

figure

1) A form or shape.

2) A symbol representing a number.

finite

Counting which can come to an end.

foot

A unit of linear measurement, one of the basic units of the imperial system. Originally the length of a man's foot.

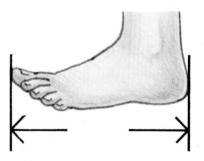

force

That which produces or tends to produce changes in the motion of bodies on which it acts. It is measured by the acceleration it causes in a body of known mass. The SI unit of force is the newton (N).

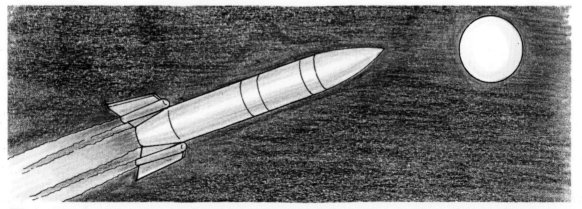

formula

A rule or principle for solving mathematical problems usually expressed in algebraic symbols. The formula for finding the area of an oblong is length×breadth (l×b).

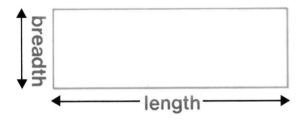

fraction

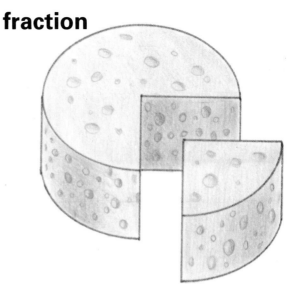

A part of a unit. It can be expressed in the form of a "vulgar" (common) fraction or a decimal fraction; e.g. ¼ or 0.25.

furlong

A unit of linear measurement. There are eight furlongs to a mile. Originally, the distance an ox-team could plough before they needed to rest.

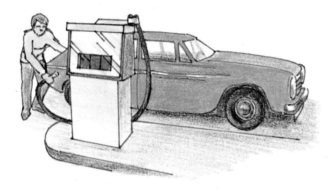

gallon

British imperial standard measure of capacity equal to 160 fluid ounces.
1 imperial gallon=4 quarts=8 pints. It is being replaced by the metric measure.
1 imperial gallon=4.55 litres (approximately).
1 U.S. gallon=128 U.S fluid ounces.

gauge

1) An instrument used to measure the diameter of wire, thickness of metals, etc.

2) To gauge means to make an estimate.

geometry

A branch of mathematics that studies the properties and relationships of lines, surfaces and solids in space.

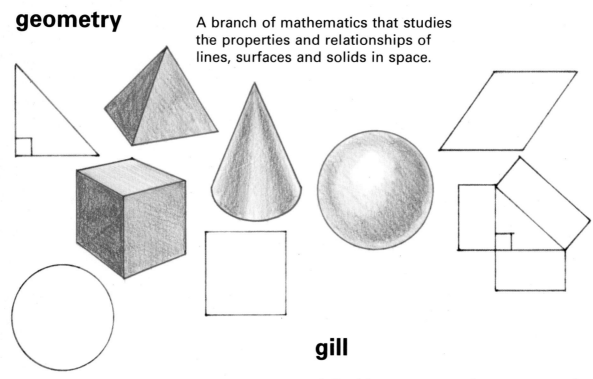

gill

A liquid measure equal to a quarter of a pint.

gradient

A measure of the steepness of slope. A gradient of 1 in 7 means that for every seven units you move in a horizontal direction while going up such a hill, you are one unit higher in a vertical direction.

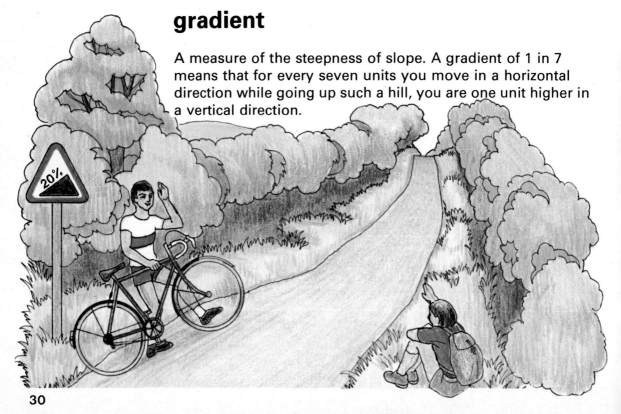

grain

A unit of weight originally equal to a grain of wheat from the middle of the wheat-ear and probably the first standard unit of weight. It was the smallest British weight, and was also used by chemists as a unit of measurement for drugs. Now in the process of being replaced by milligrams.

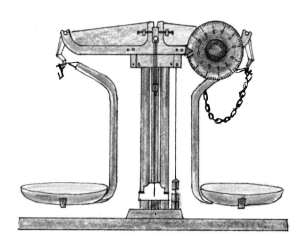

gram (gramme)

A unit of mass in the metric system. 1000 grams=1 kilogram. Shorter spelling of gram is favoured rather than gramme by SI usage. SI symbol g.

graph

A chart or diagram to show facts clearly. A line graph is often used to show changes and/or to compare values. The figures are recorded on the graph as points and each point is joined to the next by a straight line or curve.

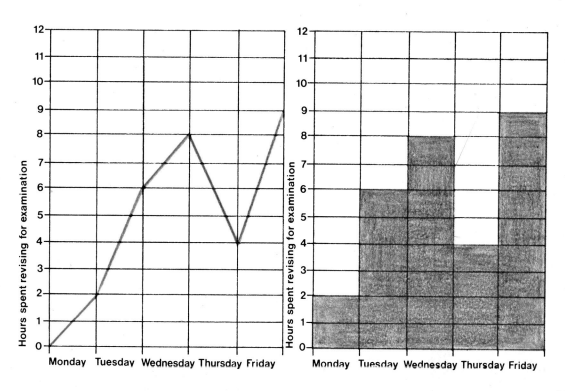

gravity

The force that draws towards a centre of attraction; e.g. the attraction of the earth's mass for bodies on or near its surface. Sir Isaac Newton (1642–1727), having watched apples falling from a tree and wondered why they fall to earth instead of floating off into space, found that the earth operates like a large magnet, attracting all bodies within a certain distance from its surface.

greater than

A term used to show which is the larger value of two numbers. The symbol is >, as in 5>2.

Greenwich meridian

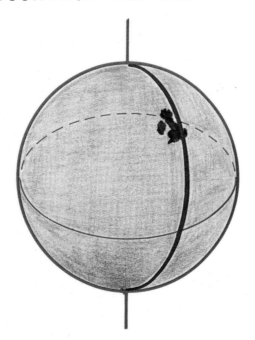

The line of longitude (meridian) taken as zero, which passes through Greenwich in south London and on which the British Royal Observatory stood. Since local time varies according to longitude, Greenwich mean-time (the time of day at the Greenwich meridian as shown by a clock which reads 12 noon when the sun is at its greatest height) is used as a standard all over the world.

gross

Twelve dozen, i.e. 12×12, or 144.

hand

A unit of measurement used for an animal's height; approximately four inches. Originally the width of a hand.

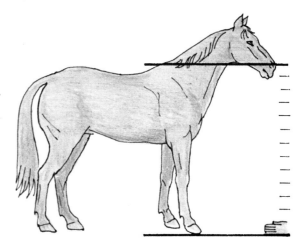

hectare

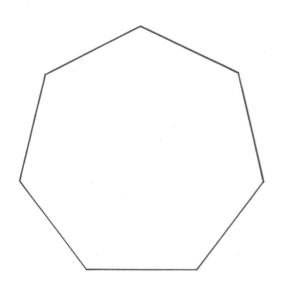

Metric unit used for measuring areas of land such as fields and parks.
1 hectare = 10,000 square metres.
SI symbol ha.

height

The distance from the bottom or base of an object to its top. Altitude.

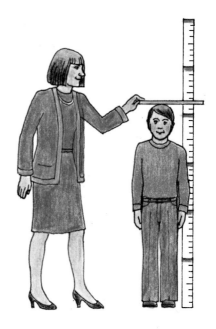

heptagon

A plane (flat) figure with seven sides and seven angles.

hexagon

A plane (flat) figure with six sides and six angles. A regular hexagon has all sides and angles equal.

hexahedron

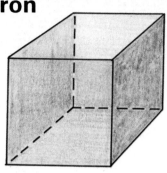

A solid made of six faces. A regular hexahedron has six identical square faces with three squares meeting at each corner. Another name for this solid is a cube.

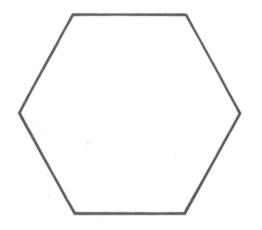

histogram

A pictorial graph.

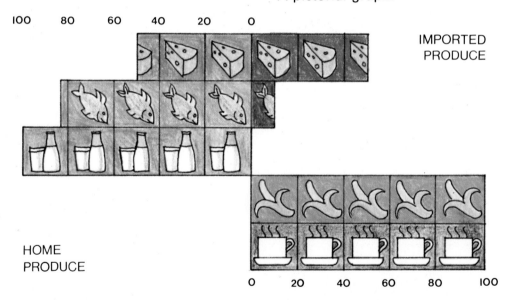

horizontal

Parallel to the horizon or to flat ground. Measured in such a plane.

hour

Unit of time equal to 60 minutes.

hourglass

An instrument for measuring intervals of time by letting sand fall from the upper to the lower of two glass bulbs joined neck to neck. The amount of time this takes is determined by the width of the passage from one bulb to the other, and by the quantity of sand used. Hourglasses are commonly used today as eggtimers.

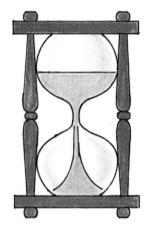

hundredweight

A measure of weight equal to 112 lb. Originally 100 lb. Symbol cwt.

hyperbola

The curve made when a plane cuts both branches of a cone. The outline of a cooling-tower shows the sides of the tower as hyperbolic curves.

hypotenuse

The longest side of a right-angled triangle; i.e. the side opposite the right angle. The length of one unknown side of a right-angled triangle can be calculated, if the length of the other two is known, by using the Pythagorean theorem.

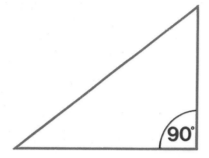

icosahedron

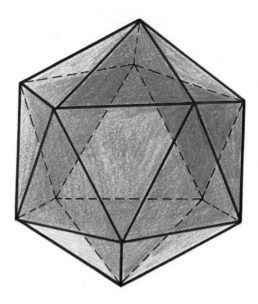

A twenty-sided solid. A regular icosahedron has twenty triangular faces and five triangles meeting at each corner.

imperial measures

The weights and measures whose standards were fixed by law for the United Kingdom. The measures used in this system come from a variety of origins and were mostly standardised early in the nineteenth century. They are still used in Britain and many English-speaking countries, although they are in the process of being replaced by the metric system.

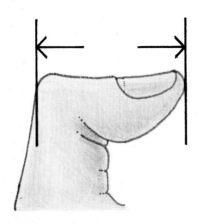

inch

A measure of length in the imperial system, equal to one twelfth of a foot. Originally it was the width of a human thumb.

index

The figure showing the power to which a quantity is to be raised; e.g. in the equation $2^3=8$, the index is 3.

$$3^2 = 3 \times 3$$
$$3^3 = 3 \times 3 \times 3$$
$$3^4 = 3 \times 3 \times 3 \times 3$$

$1\,cm \times 1\,cm = 1\,cm^2$

infinite

When counting can never come to an end; e.g. the set of even numbers is infinite.

System of integers

$-4\ -3\ -2\ -1\ 0\ +1\ +2\ +3\ +4$

Negative integers   Positive integers

integer

One of the set of numbers which includes zero and all whole numbers, with their negative numbers. Positive integers, all numbers greater than 0, are given a plus sign $(+)$; negative integers, those less than 0, carry a minus sign $(-)$.

interest

Extra payment for the use of money. The sum of money on loan is called the principal, and the extra is called interest. Interest is always reckoned as a percentage. There are two kinds of interest: simple and compound. Simple interest is a percentage of the principal. Compound interest is the interest added to the principal to make a new principal for the next period. If you invest money in a building society, you earn interest and your money increases. If, on the other hand, you borrow money from a building society to get a mortgage on a house, you will pay interest as well as paying back the money you borrowed (the principal).

international system of units

The modern version of the metric system of units, designed to be used in all countries of the world. Symbol SI.

children who bought fruit

intersection of sets

Where there are members common to two or more sets, those sets are said to intersect.

irrational number

A real number which cannot be expressed as a whole number or a fraction.

isosceles triangle

A triangle with two equal sides and therefore two equal angles.

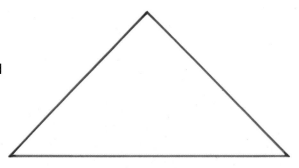

Kelvin scale

A temperature scale (sometimes called the absolute scale) which has a degree equal to the centigrade degree but with a zero equal to −273°C. The freezing-point of water on this scale is therefore 273°K. Named after the British physicist Kelvin (1824–1907) who first used this scale. The kelvin is the base unit of temperature in the SI metric system (SI symbol K).

kilogram

The base unit of mass in the metric system (*kilo* is Greek for one thousand). The shorter spelling of kilogram is favoured rather than kilogramme. SI symbol kg.

kilometre

A metric unit of length (SI symbol km) equal to 1000 metres. A kilometre is shorter than a mile; roughly 8 km=5 miles.

kilowatt

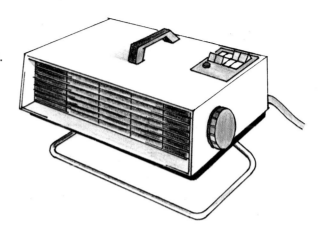

A unit of power equal to 1000 watts. SI symbol kW.

knot

A unit of speed equal to one nautical mile (6,080 ft.) per hour. 13 knots=15 miles per hour. The name comes from the old practice of letting a rope, knotted at regular intervals, run out as the ship moved – the speed could be calculated by the number of knots run out in a certain time.

latitude

Lines of latitude are a series of imaginary circles drawn round the earth. The circle halfway between the North and South Poles is called the equator and is 0°. The Poles are 90°N and 90°S. Points that are on the same latitude are the same distance north or south from the equator.

length

The longest measurement of anything from end to end; or any linear measurement between two points in space or time.

line

A set of points joined together which can be straight or curved. A line has length but no thickness; i.e. it has only one dimension.

line segment

A part of a line.

1 LITRE = 1,000 MILLILITRES
1 LITRE = 1,000 CUBIC CENTIMETRES

logarithm

The power to which a fixed number b (the base, usually 10) must be raised to be equal to the number a in question. If $a=b^n$, n is the logarithm of a to base b; e.g. the logarithm of 100 to base 10 is 2. Logarithms are arranged in tables and make complicated calculations easier by changing the operations of multiplication and division to addition and subtraction. If you work out $10^2 \times 10^5$ you will see that the answer is 10^7, in other words the powers are added for multiplication of numbers with the same base.

less than

A term used to show which has the small value of two numbers. The symbol is <, as in 2<4.

$$5 < 7$$

linear measure

Measurement by length.

litre

The metric unit of measurement used for capacity or volume of liquids and gases, equal to a cubic decimetre. SI symbol l or litre. 1 litre=approximately 1.76 pints.

$3^3 = 3 \times 3 \times 3 = 27$
logarithm of 27 is 3
$3^4 = 3 \times 3 \times 3 \times 3 = 81$
logarithm of 81 is 4
to multiply 27×81 is the same as multiplying $3^3 \times 3^4$
$27 \times 81 = 2187$
the logarithm of 2187 is 7
the logs of 27 and 81 i.e. 3 and 4 were added to get 7

longitude

Lines of longitude are half meridians from one pole to another. They are a series of imaginary semi-circles drawn on the surface of the earth. When this system was originated, the zero line was set to go through Greenwich, England, and other longitudes are measured in degrees west or east of this.

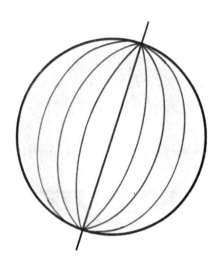

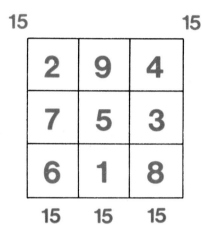

"magic" square

A square chart where the numbers in any row, column or diagonal always add up to the same sum.

magnetic pole

The point on the earth's surface at which the magnetic lines of force meet. The magnetic north pole is not at the North Pole but somewhere in Hudson Bay. Similarly, the magnetic south pole is not at the South Pole but in the Antarctic Ocean. The magnetic poles are variable.

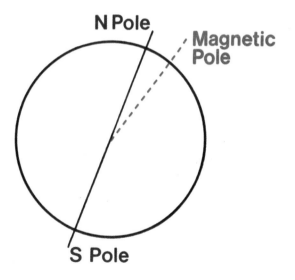

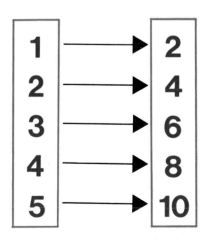

mapping

A matching operation between two sets in which each member of one set is related to a member of the second set. Every member of the first set must have a partner in the second, though several can share the same partner.

mass

The quantity or bulk of matter. The mass of an object, such as a man, is the same on the earth or in space, but his weight is not the same because gravity plays an important part in determining weight. The mass of an object is not affected by the conditions around it. The SI unit for mass is the kilogram (symbol kg).

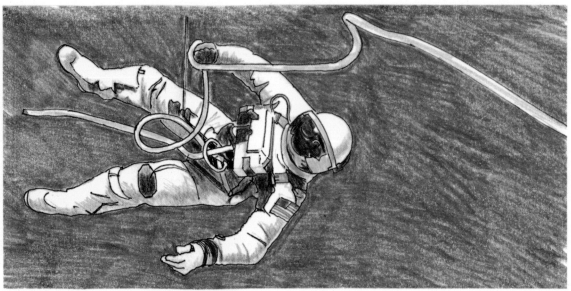

matching

One to one correspondence when each member of one set is put in relationship with one member of another set.

mathematics

The science of number and space. Methods of finding unknown quantities from known quantities. Pure mathematics includes geometry, algebra, trigonometry, etc. Applied mathematics puts the same techniques and formulae into practice in relation to the physical world and is important in the fields of physics, engineering, astronomy, etc.

$3 + 9 = 12 \qquad 10 \div 2 = 5$

$5 \times 3 = 15 \qquad 6 - 2 = 4$

$2 + \square = 10 \qquad 5 + a = 10$

$2^2 = 4 \qquad \frac{1}{2} + \frac{3}{4} = 1\frac{1}{4}$

$(3 + 4) \times 5 = 7 \times 5 = 35$

$1, 2, 4, 7, 11, 16, \square$

$2 \cdot 6 + 7 \cdot 4 = 10$

$x + y = 14$

mean-time

Clock time. The length of the clock day is the average of all the solar days in the year, divided into uniform hours, minutes and seconds. For most of the year there is a slight difference between clock time and the time according to a sundial.

measure

Find the size of anything and express this in terms of standard units. Most countries of the world use the metric system as standard units of measurement.

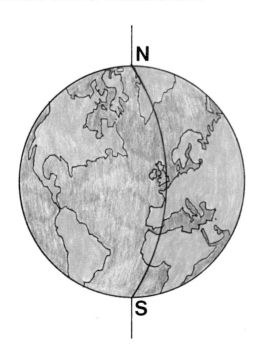

meridian

A meridian is any imaginary line passing through both the North and South Poles to make a circle round the earth. Each meridian has a longitude measured in degrees to show its position in relation to the standard or prime meridian (0°) which passes through Greenwich, England.

metre

The metric base unit of length (SI symbol m).

metric system

A decimal system first adopted in France in 1795. The standard units are based on measurements of the earth and water. A quadrant (one quarter of a meridian) was divided into ten million equal parts. The length of one of these was called a metre. Its length was measured between two scratches on a platinum bar which is kept in a vault in Paris. The metric unit of mass is a kilogram. In 1964 the kilogram was defined as the mass of one cubic decimetre (one litre) of water.

mile

An imperial measure of length equal to 1,760 yards. Originally a Roman measure of 1,000 paces. A geographical mile (used in measurements of the earth's surface) is equal to 6,082.66 feet and is one minute of equatorial longitude. A nautical mile (used by ships) is equal to 6,080 feet.

milligram

A unit of mass. 1000 milligrams= 1 gram (*mille* means a thousand in Latin). SI symbol mg.

millilitre

A unit of volume. 1000 millilitres= 1 litre. SI symbol ml.

millimetre

A unit of length, 1000 millimetres=1 metre. SI symbol mm.

minute

1) A unit of time. 60 minutes=1 hour. Symbol min.

2) In geometry a unit equal to a sixtieth of a degree. Symbol '.

month

One of the twelve divisions of the year, roughly corresponding to the time taken for the moon to rotate once round the earth (about 29½ days).

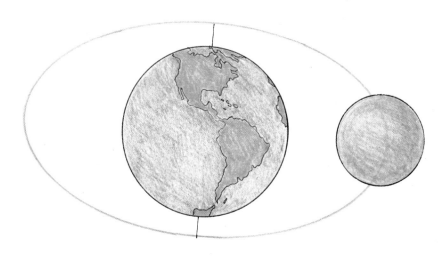

multiplication

To add a number to itself a certain number of times; the addition of equal sets. The sign × means "multiplied by" or "times". 5×3 shows that three fives (or five threes) are added together.

5 × 3 = 15
multiplicand x multiplier = product

multiplication square

1	2	3	4	5	6
2	4	6	8	10	12
3	6	9	12	15	18
4	8	12	16	20	24
5	10	15	20	25	30
6	12	18	24	30	36

natural number

One of the numbers used for counting; i.e. whole numbers over 0. The oldest and the smallest of the number systems.

navigation

Steering a course in any kind of craft. (*See* dead reckoning.)

net

A shape or pattern drawn in such a way on paper that it can be folded to make a solid.

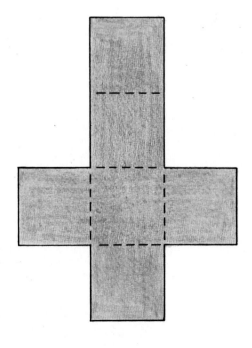

nought

See zero.

number

The symbol or numeral which records quantity and answers the question "how many?" Where there are separate sets, for example one containing three apples, one three horses and one three trees, the property they all have in common is the abstract one of number, in this case the number three.

number line

A way of picturing the abstract idea of numbers as points on a line. The points are marked off at equal distances and numbered in order. The line can be extended indefinitely in either direction, positive numbers increasing to the right and negative numbers decreasing to the left.

number sentence

A statement in mathematics; e.g. 7+5=12.

numeral

A symbol indicating a number; e.g. 5.

numerator

The figure above the line in a vulgar fraction showing the number of fractional parts. The figure below the line is called the denominator.

3/4 ←3 is the numerator

oblique

Slanting; at an angle; not parallel with the horizontal or vertical plane.

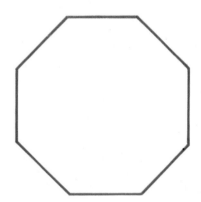

oblong

A rectangle which is not a square; a four-sided plane figure with opposite sides equal and four right angles.

octagon

An eight-sided plane (flat) figure. A regular octagon has eight equal sides and eight equal angles.

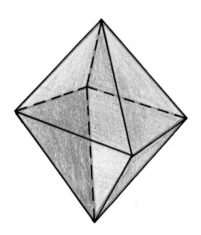

octahedron

A solid with eight faces. A regular octahedron has eight triangular faces and four triangles meeting at each corner.

odd number

Any positive whole number of which two is not a factor; i.e. any number which cannot be divided exactly by two.

1 3 5 7 9
11 23 35 47 59
61 73 85 97 109

odometer

A digital instrument which records distance travelled, such as in a car.

one to one correspondence

A term showing that one member of a set matches only one member of another set.

open sentence

A mathematical sentence puzzle with an unknown element marked by a placeholder. The truth set which replaces the placeholder makes it into a closed (true and complete) sentence.

$10 + \square = 32$

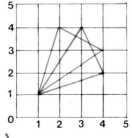

ordered pair

$(1,1)(1,1)$
$(2,4)(4,2)$
$(4,3)(3,4)$

A set containing two elements in which the order is important (such as co-ordinates). A different set is produced if the order is reversed. The relation of ordered pairs can be shown on a graph. They are written in parentheses; e.g. (2,4).

ordinal number

A number that marks the order of a member in a set; e.g. first, twenty-seventh.

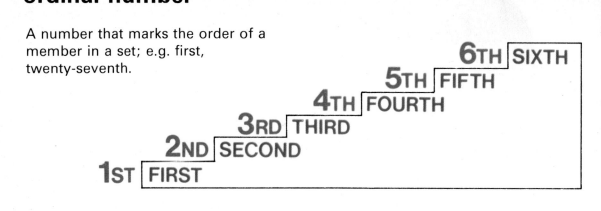

ordinality

Arrangement in sets or ranks according to some definite successive order; e.g. putting coins in order of value:
50p 10p 5p 2p 1p ½p

ounce

1) A unit of mass in the imperial system. Symbol oz.

2) A unit of liquid measure (fluid ounce) in the imperial system.

16 ounces = 1 pound

Troy weight
12 ounces = 1 pound

palm

A measure used in Ancient Egypt equal to the width of four fingers, or one seventh of a cubit.

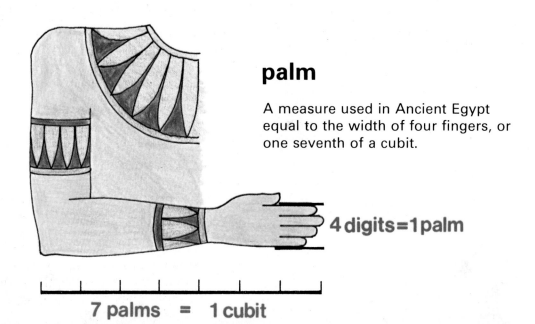

parabola

An open curve made by cutting a cone with a plane parallel to one of its sides. The path of a stone thrown through the air is a parabola.

parallel lines

Lines in the same plane that are equally distant from each other at all points and never meet.

parallelogram

A four-sided plane (flat) figure whose opposite sides are equal and parallel; i.e. rectangle, rhombus, square, rhomboid.

parentheses

The symbols ( ) which are used to show the grouping or order of operations. The operation inside the parentheses is completed first where possible.

$$(3+4) \times 5$$
$$= 7 \times 5$$
$$= 35$$

partitioning

Separating a set into two or more subsets.

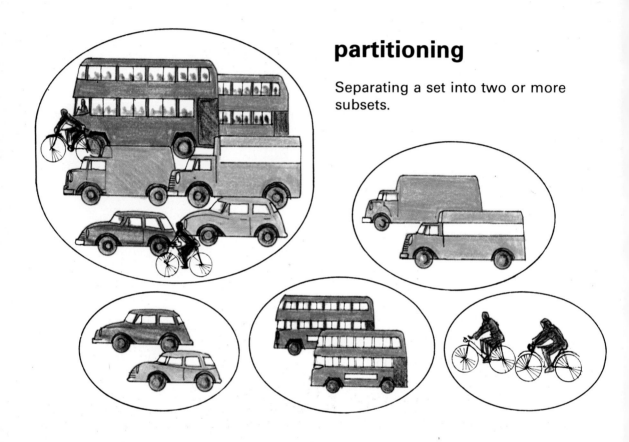

peck

An old measure of capacity which was used for dry goods, equal to a quarter of a bushel.

pence

The British decimal currency system has the pound (£) as the basic unit, which is divided into 100 smaller units called pence (100p=£1), or sometimes "new pence" to differentiate them from the former system, replaced in 1971, of pounds, shillings and pence (£ s d).

pennyweight *See* troy weight.

pentagon

A plane (flat) five-sided figure. A regular pentagon has five equal sides and angles.

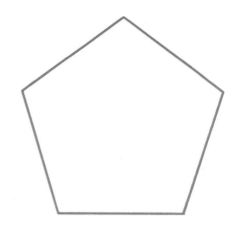

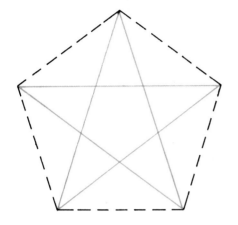

pentagram

A five-pointed star made by the five diagonals of a pentagon.

percentage

A measure of proportion in terms of amount or quantity per hundred. Interest on savings is paid on this basis and some people, especially salesmen, are paid "on commission" whereby they get a fixed percentage of the value of their sales. If you are paid a five per cent (5%) commission and you sell £200 worth of goods, £10 of that will be yours.

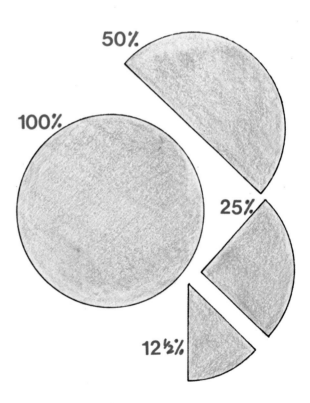

perch

An old measure, no longer used, equal to 5½ yards in the imperial system.

perimeter

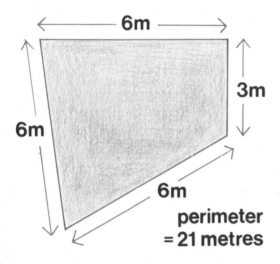

The line or set of lines forming the boundary of any closed figure. The length of this boundary.

perpendicular

Vertical or exactly upright; at right angles to a given surface.

pi

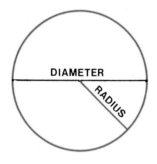

The circumference of a circle is about 3.14159 times the length of its diameter. As the exact number cannot be written as a fraction or decimal but is the same for any circle, the Greek letter π (pi) is used to stand for it.

pictogram

A diagram which shows information in pictures or symbols for quick understanding.

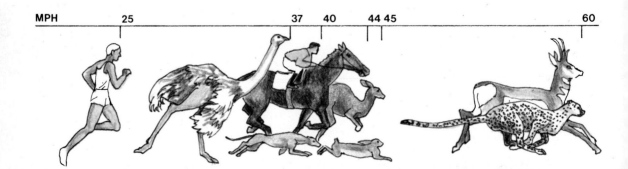

pie chart

A circular diagram used to show information. A full circle is 360° and this can be divided into sectors by using a protractor.

36 cars in a car park

how many new cars?

pint

Unit of capacity, mostly used for liquids, in the imperial system of measures. It is in the process of being replaced by the metric unit, the litre. In British imperial measures
1 pint = 20 fluid ounces. In U.S. measures
1 U.S. pint = 16 U.S. fluid ounces.

$6 + \square = 10$
$6 + ? = 10$
$6 + \triangle = 10$
$6 + \_\_ = 10$

placeholder

A symbol used for the unknown quantity in an open mathematical sentence — the set of numbers which replaces this symbol to make a true and complete sentence is called the truth set.

place value

The value of a symbol which depends on its position in a number system; e.g. the value of the symbol 3 is different in these two numbers, 3 and 300.

$$250 = (2 \times 100) + (5 \times 10) + 0$$
$$25 = (2 \times 10) + 5$$
$$2 \cdot 5 = 2 + \cdot 5$$

plane

A surface that has two dimensions. As two points are needed to define a line, three or more points are needed to define a plane; e.g. a square is a plane shape.

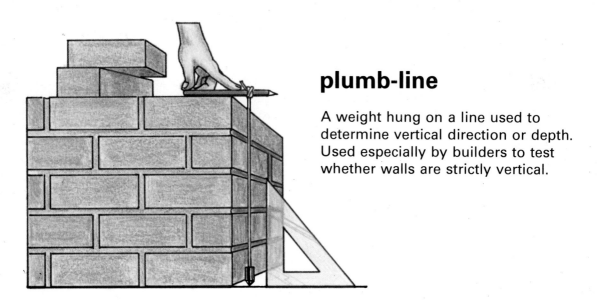

plumb-line

A weight hung on a line used to determine vertical direction or depth. Used especially by builders to test whether walls are strictly vertical.

polygon

A closed plane (flat) figure whose sides (usually five or more) are straight lines. If it has all equal sides and all equal angles it is a regular polygon.

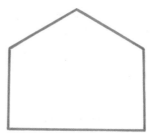

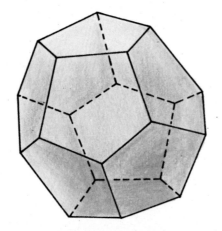

polyhedron

A solid with many faces, usually six or more. There are five regular polyhedrons (that is, the faces are the same regular polygons with the same number of polygons at each corner) — tetrahedron, hexahedron, octahedron, dodecahedron and icosahedron.

post meridiem

The Latin words for "after midday". The abbreviation p.m. is used to indicate times between midday and midnight.

pound

1) Basic unit of British currency (symbol £).

2) A unit of mass in the imperial system, equal to 16 ounces (symbol lb).

power

$$2^2 = 2 \times 2$$
$$2^5 = 2 \times 2 \times 2 \times 2 \times 2$$

The number of times a number is to be multiplied by itself, represented by the index number; e.g. $y^4 = y \times y \times y \times y$ (y to the power of four).

prime number

A natural number that cannot be made by multiplying smaller numbers together. It can only be divided evenly by itself and one.

1 2 3 5 7 11 13

prism

A solid whose two end faces are equal in area and shape and in parallel planes. The side faces that connect them are parallelograms.

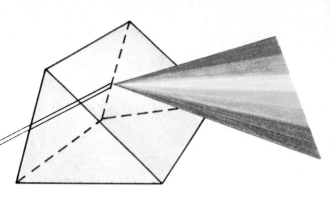

probability

A branch of mathematics dealing with calculations of how likely events are to happen. For example, when a coin is tossed there are two possible results, heads or tails, so the probability of the result being heads is 1 in 2, or ½.

product

The total obtained by multiplying two or more numbers together.

5 × 6 = 30
30 is the product

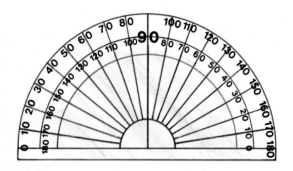

protractor

A circular or semi-circular instrument marked for measuring angles.

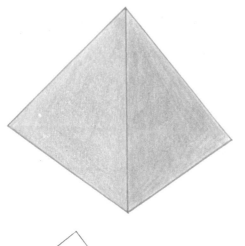

pyramid

A solid whose base is a straight-lined figure and whose sides are triangles which meet in a point called the apex.

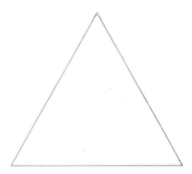

Pythagorean theorem

Pythagoras, a Greek mathematician, discovered about 2500 years ago that, in a right-angled triangle, the square on the side of the hypotenuse is equal to the sum of the squares on the other two sides. Using this formula, it is possible to calculate the length of any one side if the other two are known.

$$3^2 + 4^2 = 5^2$$
$$9 + 16 = 25$$

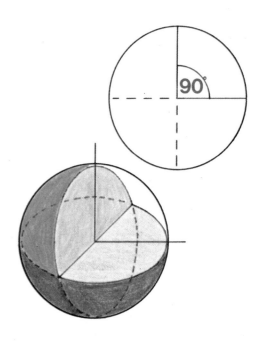

quadrant

A quarter of a circle's circumference or area as cut by two diameters at right angles or a quarter of a sphere as cut by two planes intersecting at right angles.

quadrilateral

A four-sided plane figure.

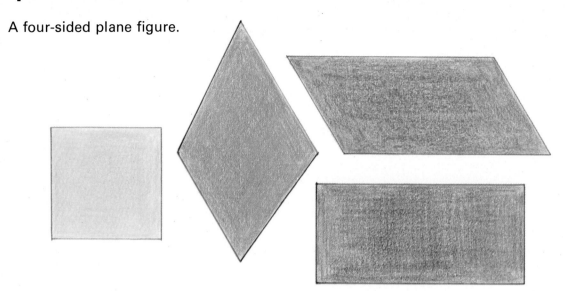

quart

Unit of capacity in the imperial system. In British measures 1 quart=2 pints=40 fluid ounces. In U.S. measures 1 U.S. quart=2 U.S. pints=32 U.S. fluid ounces.

quire

24 sheets of paper of equal size.

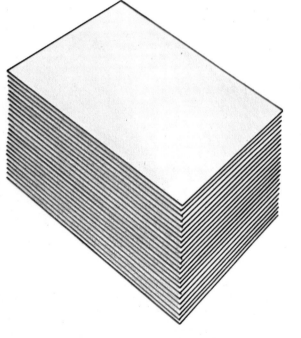

quotient

The number obtained by the division of one number by another; i.e. the result of a division sum.

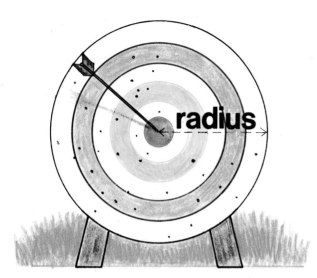

radius

A straight line from the centre of a circle to any point on its circumference. The length of such a line.

ratio

A ratio compares one quantity with another (symbol :); e.g. a boy 1.5 m tall casts a shadow 3 m long, the ratio of his height to his shadow's length is 1.5:3 or 1:2.

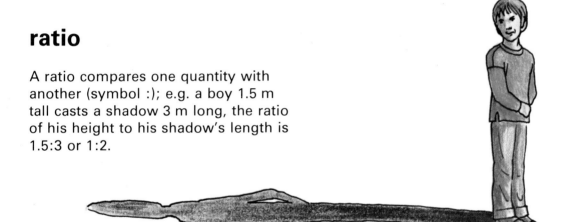

rational number

A number which can be expressed as a fraction.

real numbers

An expanded number system in which there is a number for every possible point on the number line. This includes integers (whole numbers), fractions, all rational and irrational numbers.

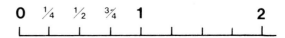

rectangle

A four-sided plane (flat) figure with four right angles and the opposite sides parallel and equal in length.

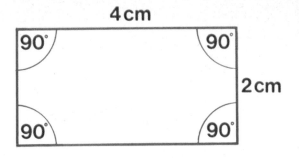

rectangle number

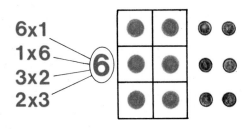

Any number that can be shown as a rectangular pattern of dots; e.g. 6, 8, 10. Every rectangular number is the product of smaller numbers; prime numbers therefore cannot be rectangle numbers.

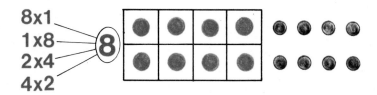

reflection

The image given back from a mirror or other reflecting surface. The image is "laterally inverted" or turned round from the original; i.e. what looks like the left ear on the face looking out from the mirror is actually your right ear. It also appears to be at the same distance behind the mirror as you are in front of it.

region

The area within a closed curve.

regular

Shapes that have equal sides and equal angles are regular. Polygons that have equal sides and equal angles are called regular polygons. A regular polygon fits exactly into a circle. Solids with faces that are regular polygons and the same number of polygons at each corner are regular solids. There are five regular solids: tetrahedron, hexahedron, octahedron, dodecahedron and icosahedron.

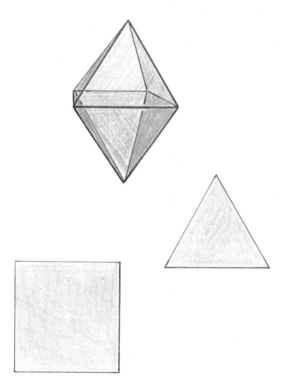

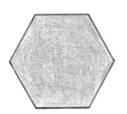

relation

How members of one set compare with members of another; e.g. "greater than", "less than", etc.

rhomboid

A parallelogram whose angles are not right angles. It is like a rhombus but only the opposite sides are of equal length.

rhombus

A four-sided plane (flat) figure whose sides are all equal but whose angles are not right angles.

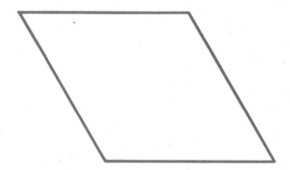

right angle

An angle of 90°. The simplest way of drawing a right angle is to use a set square.

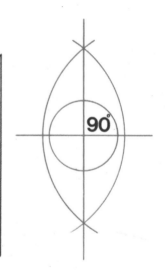

right-angled triangle

A triangle with a right angle. The relative lengths of the sides are expressed in the Pythagorean theorem.

rod

A linear measure equal to 5½ yards. One of the earliest standard measures. The Roman rod was 10 feet and in the Middle Ages the rod was the total length of 16 men's left feet.

Roman numerals

The Roman system of writing numbers. Today the dials of clocks often carry Roman numerals and books are often numbered as *Vol I* and *Vol II*, etc.

```
1  2  3  4  5  6   7   8    9  10
I  II III IV V VI VII VIII IX  X

50 100 500 1000
L   C   D   M
```

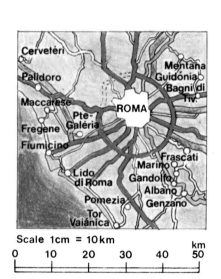

rood

An old measure of land equal to ¼ acre.

scale

The ratio of a drawing or model's size to the size of the object it represents; e.g. 1 cm = 1 m which means that 1 centimetre represents one metre.

score

A term for the number twenty, not much used now.

second

1) A unit of time, the sixtieth part of a minute of time (symbol s).

2) A unit in the measurement of angles, the sixtieth part of a minute (symbol ″).

3) An ordinal number, next after first.

section

A plan of the inside of a solid as if it had been cut through by a plane.

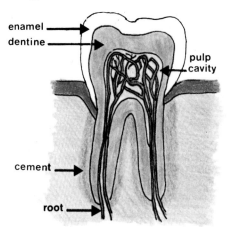

longitudinal section of a human tooth

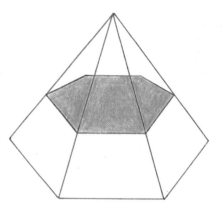

sector

The part of a circle bounded by two radii and the arc between them.

segment

A part of a curve, a solid, a plane (flat) shape or a line.

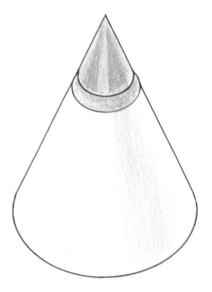

semicircle

Half a circle bounded by a diameter and half the circumference (*semi* is the Latin word for half).

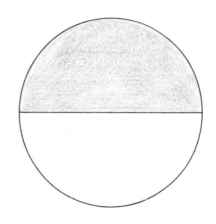

sequence

A set of numbers or objects arranged according to some principle of pattern or order.

1, 2, 4, 7, 11, 16, ☐
What is the next number in this sequence?

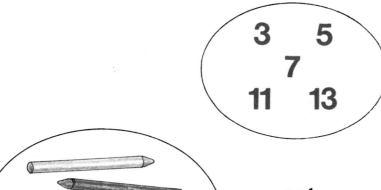

set

Any collection of objects, known as members or elements of the set, with one or more specified property in common. It can be, for example, a set of numbers or the set of children with brown eyes.

67

set square

A triangular instrument with one angle of 90°, used for measuring right angles.

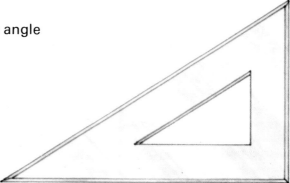

sextant

An instrument used to measure the angle between the lines joining the user to two distant objects; e.g. the horizon and the sun, moon or a star so as to measure their altitude. From this information, and with the aid of a chronometer and the nautical almanac which give the time and the positions of the planets at Greenwich, a navigator can calculate his latitude and longitude.

SI

The symbol for the international system of units.

slide rule

An instrument consisting of two inter-connecting rulers. Simple slide rules are used for addition and subtraction. Others are designed for multiplication and division.

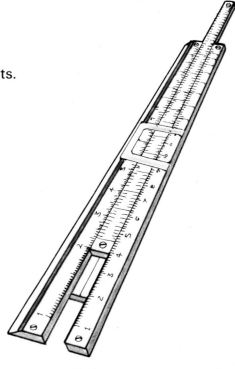

solid shape

A body with three dimensions, length, breadth and height.

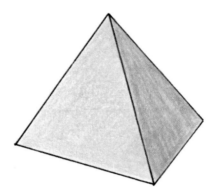
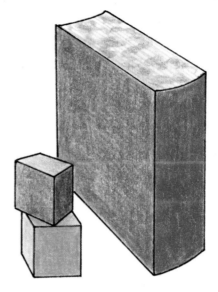

sphere

A closed solid bounded by a surface on which every point is equidistant from a central point within the solid; e.g. the globe or a ball.

spiral

Coiled in a plane (two-dimensional) or as if round a cylinder or cone (three-dimensional); e.g. a spring or corkscrew.

spirit level

An instrument used, especially by builders, to test if surfaces are horizontal. It has a short glass tube of alcohol with an air bubble in it which rests halfway along the tube if the surface is horizontal.

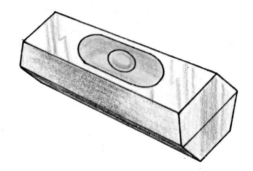

square

A four-sided plane (flat) figure with all its sides the same length and all its angles right angles.

square of a number

$2^2 = 4$ $4^2 = 16$
$3^2 = 9$ $5^2 = 25$

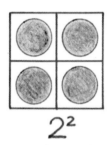 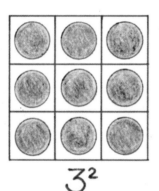

The product of multiplying a number by itself. Six squared is written as 6^2 and means 6×6, i.e. the square of six is 36. The small ^{2} written in the upper right-hand corner is the index number or power, and shows that the number has to be multiplied by itself.

square root

A number which multiplied by itself produces the given number. For example, $4 \times 4 = 16$ so the square root of 16 is 4. The symbol for square root is $\sqrt{}$, e.g. $\sqrt{9} = 3$.

$5^2 = 25 \therefore \sqrt{25} = 5$
because $5^2 = 5 \times 5 = 25$

$\sqrt{9} = 3$ $\sqrt{4} = 2$
$\sqrt{25} = 5$

standard measures

Units of measurement fixed by the government of a country. Most countries of the world use the metric system as the standard system of measurement but the imperial system is still used in many English-speaking countries. There are physical standards according to which units are defined; e.g. the yard is defined as the distance between two fine lines engraved on gold studs sunk in a particular bronze bar, cast in 1845.

statistics

1) A collection of numerical facts which are arranged in order.

2) A branch of mathematics which studies the collection of numerical facts, in such areas as population figures, average earnings, etc.

population

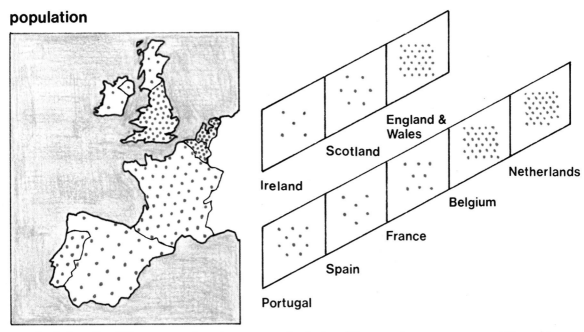

each dot = 1 million people

each dot = 10 people per square kilometre

stone

A unit of mass in the imperial system equal to 14 pounds.

subset

A set contained within a larger set; e.g. dogs are a subset of the set of animals.

subtraction

The operation of finding the difference or taking away. The sign − (minus) is used to show the operation. At first sight it may seem that a problem such as 3−6, for instance, would not be possible but if the negative integers are used every subtraction operation has an answer.

$$20 - 7 = 13$$
$$1\tfrac{7}{8} - \tfrac{3}{8} = 1\tfrac{1}{2}$$
$$2 \cdot 45 - 1 \cdot 04 = 1 \cdot 41$$

sundial

An instrument that measures time by the shadow of a pointer (gnomon) falling on the time divisions marked on the dial, the shadow moving as the position of the sun in the sky changes.

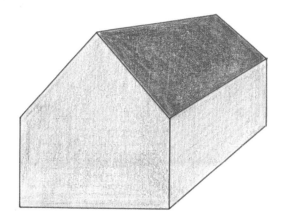

surface

One of the faces of a solid. It has length and breadth but no thickness.

surveying

The science of measuring land surfaces and representing them on a diagram or map. Surveyors also inspect the condition of houses.

symbols

Symbols are a short way of writing mathematical statements, instructions etc., as a sort of code; e.g. + −.

symmetry

When each side of a line (real or imaginary) is a reflection of the other. Any line which divides an object so as to have this effect is called an axis (or line) of symmetry. There can be more than one line of symmetry; e.g. a square has four.

tally

A primitive method of counting by making a notch on, for example, a piece of wood for each object. If sheep were counted like this, each notch would represent one sheep, so the total of each should correspond.

tangram

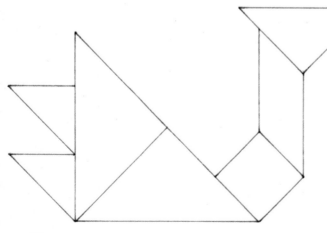

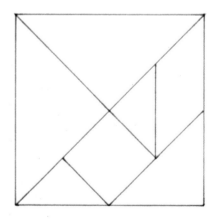

A square cut into seven pieces in a special way that can be rearranged to form various other shapes. Originally a Chinese puzzle.

tessellation

Shapes fitted together to form a patterned mosaic-like surface. A tessellation of all one kind of regular polygons is called a regular tessellation.

tetrahedron

A four-sided solid. A regular tetrahedron is made of four triangles of the same size and shape, with three triangles at each corner.

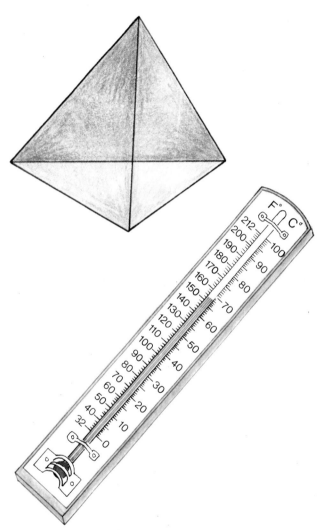

theodolite

An instrument used in surveying to measure horizontal and vertical angles. Simply, it is a telescope fixed above a horizontal circular scale marked in degrees, rather like a protractor. The telescope can also be pointed up or down and the angles read off a vertical scale.

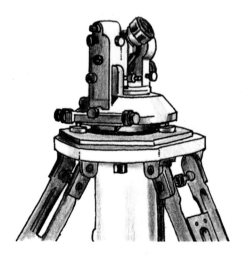

thermometer

An instrument for measuring temperature. One of the most common types has a small thread of mercury inside and the transparent container is marked according to where the mercury will expand to at which temperatures. They can use any of the three scales of temperature, Celsius (centigrade), Fahrenheit or Kelvin.

three-dimensional

Having height, breadth and length or depth which can be measured; e.g. a room. Solids are three-dimensional, planes only two-dimensional.

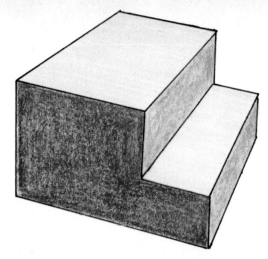

time

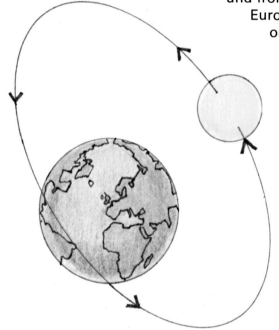

A period during which things happen. Clocks show the time. In Britain the twenty-four hours are divided into two periods of twelve hours – from midnight to midday (a.m.) and from midday to midnight (p.m.). The Europeans use a 24-hour clock; e.g. half-past one in the afternoon is 1.30 p.m. in Britain but 13.30 hours in Europe. The number before the point shows the number of hours, the number after the point shows the number of minutes past that hour. As the earth makes one complete spin round on its axis each day, from west to east, half of it is in sunlight while the other half is in the dark, and at different degrees of longitude the day is at different stages. As there are 24 hours in a day and 360° of longitude, the time differs by one hour for each 15° you move west or east. Because of this, a navigator is able to find his longitude if he knows the time at a fixed point (*see* chronometer) and his local time (*see* sextant).

ton A unit of mass in the imperial system equal to 2,240 pounds in Britain and to 2,000 pounds in the U.S.A.

tonne

A unit of mass in the metric system which is the name for 1000 kilograms and is roughly the same weight as the imperial ton. SI symbol is t or tonne; in speech only, "metric tonne" should be used to avoid confusion with the imperial ton.

topological transformation

When a figure or object has been stretched or twisted into a different shape, but has not been cut or had another piece joined to it.

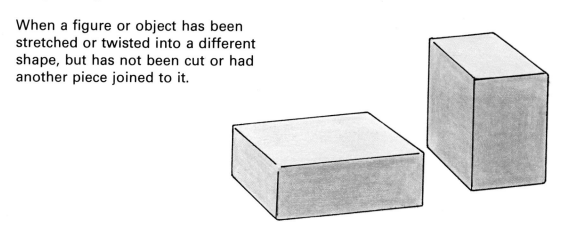

topology

The study of lines, points and the figures they make, and the way in which the area, lengths and angles can be altered.

triangle

A three-sided plane (flat) shape enclosing three angles which total 180°. Every angle in an "acute triangle" is less than 90°; in a "right-angled triangle" one angle is 90° and in an "obtuse triangle" one angle is greater than 90°.

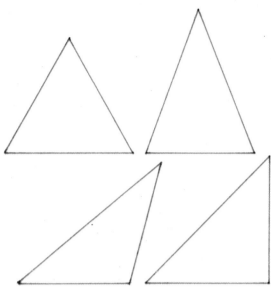

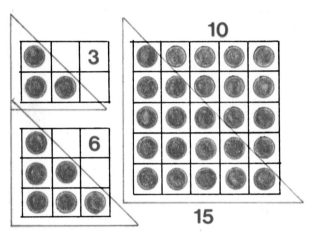

triangle number

Any number that can be shown as a triangular pattern of evenly spaced dots. It is half of a rectangle number. The first three are 1, 3, 6.

trigonometry

The branch of mathematics which deals with the sides and angles of triangles, their measurement and the relations between them. All of the angles and sides can be found when three of them are known. In applied mathematics the same principles are used to relate distances to directions.

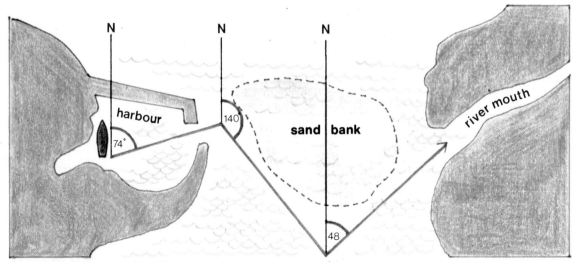

troy weight

A system of weight units used especially for precious stones and metals.

Precious metals and stones are weighed by carats of 4 grains each = 3.2 troy grains.

24 grains = 1 pennyweight
20 pennyweight = 1 troy ounce
12 troy ounces = 1 troy pound

$$2 + \square = 10$$
$$2 + 8 = 10$$
$$8 = \text{truth set}$$

truth set

The set of elements which replaces a placeholder in an open sentence to make it a true sentence.

two-dimensional

Having length and breadth that can be measured, e.g. the surface of any object.

undecagon

A plane (flat) shape with eleven sides.

union of two sets

$(2, 4, 6,) \cup (3, 5, 7,)$
$= (2, 3, 4, 5, 6, 7,)$

Putting all the elements in two sets together to form a new set which includes both. The symbol for "union of" is ∪.

unit

1) A single element.

2) A quantity or amount adopted as a standard by which other things are measured.

universal set

A set of all the elements to be considered. The universal set could be all the animals in a zoo being considered, in which case bears would be a set and polar bears a subset within this. Symbol ∪.

variable

A quantity whose value may change. In mathematics the general practice is to name the variables by the letters x, y, z but any symbol could be used. Variables are used to examine properties of all numbers or a particular set of numbers, as in x×2<8 where x<6.

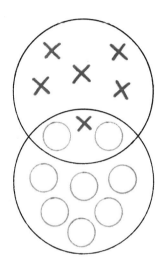

Venn diagram

Geometric shapes representing sets and showing where they intersect pictorially, to make them easier to interpret. Named after Venn, an Englishman, who first used them.

Vernier gauge

An instrument used to measure very small intervals. Invented by Pierre Vernier (1580–1637) of Brussels.

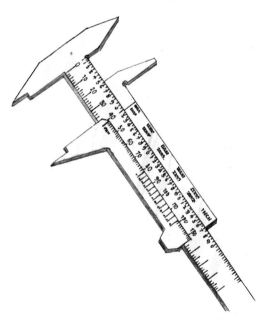

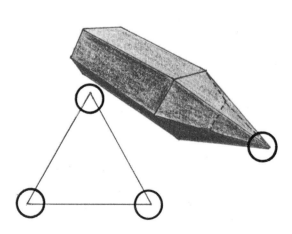

vertex

The point where lines or edges meet in a plane (flat) shape or a solid. The plural of vertex is vertices.

vertical

Upright; perpendicular (at a right angle) to the horizon.

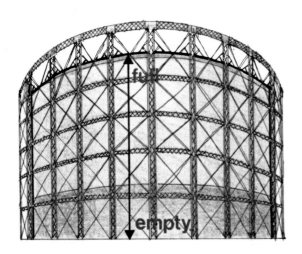

volume

The measurement of three-dimensional space taken up by a solid shape, a liquid or a gas, expressed in units of cubic measure. In everyday life the litre and millilitre are often used for the measurement of capacity or of the volume of liquids and gases (*See* capacity).

watt

A unit of power. The name is taken from James Watt (1736–1819) who invented a way by which steam engines could turn wheels.
SI symbol W.

week

A period of seven days in succession.

weigh

To find the weight of objects by balancing them against standard weights (masses) such as kilograms. To compare the weight of one object with that of another.

weight

The force produced by the pull of gravity on a given mass. On earth, where gravity has the same effect everywhere, its part can be ignored and the units of mass (the kilogram and the pound, for example) are used for weight. On the moon, which has a different gravity, an astronaut will have a different weight, although his mass will not change. (*See* mass.)

width Breadth; the distance from side to side of an object or shape.

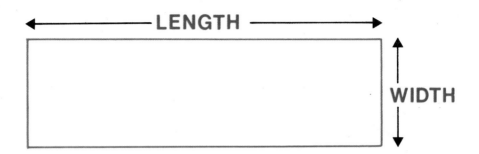

yard Unit of linear measurement in the imperial system equal to three feet or 36 inches.

year The time taken for the earth to go once around the sun (approximately 365¼ days). Because of the extra quarter day, the most commonly used calendar has a standard year of 365 days and a leap year of 366 days every four years to catch up.

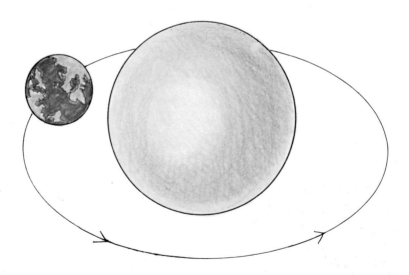

zero The symbol 0 to represent no quantity, nothing, nought. Without it the system of written numbers which is used today would not work. The first people to recognise that a symbol was needed to indicate an empty column were the Hindus of Ancient India.

The Story
of Mathematics

Mathematics began almost as soon as man appeared on earth. The earliest men and women were hunters who moved from place to place in search of food. They had few possessions and so there was no need for them to know much arithmetic. Simple counting and a method of sharing was enough.

What was more important for them was to have some knowledge of the seasons and some means of finding directions. It probably took many hundreds of years before they discovered that the stars at night are a more reliable guide to directions than the sun. For the sun's position in the sky varies as the seasons change.

The first clocks were the sun, moon and stars. During the day a hunter could roughly tell the time by noticing the length of the shadows. But to measure longer periods of time primitive people must have relied on the moon. Eventually, they discovered that there are about thirty days between one full moon and the next.

After thousands of years some hunters became farmers and shepherds. They settled in villages and owned herds of animals. It now became necessary for them to keep records of their possessions. The earliest system used was the tally system. Counting in this way lasted for a very long time and it is still used when the Roman numerals for one, two and three are written (I, II, III).

When men became farmers they needed to have a more reliable calendar than the hunter's "moon calendar" to forecast the seasons. Wise men, who were often priests, worked out a sun calendar. As a result, they frequently became the most powerful men in the land.
In time, they measured the length of the year accurately, but in order to do this they had to have a system of written numbers.

As civilisation advanced, counting, measuring and calculating became more important. About five thousand years ago the people of Ancient Egypt and Mesopotamia used a system of written numbers which are the earliest we know about.

Although these lands are far apart their number systems seem to have started in the same way. They both used simple strokes for ones and separate symbols for tens and higher numbers. Thousands of years later the Romans made strokes for the numbers one to four, but used letters for fives, tens and fifties, etc. At about the same time the Chinese used strokes for the first three numbers although they had a different sign for every other number up to ten.

The Mayas of Central America could write any number with only a dot, a stroke and an oval. However it was the Hindus of Ancient India who were the first people to recognise that they needed a symbol for zero.

In Ancient Egypt the priests became very powerful and ordered the building of huge temples and pyramids. In building these, the architects made great discoveries in the art of measurement. The farmer's rough method of measurement which he used for building a hut or house was not good enough for large-scale building. Architects had to have measures that were always the same. These measures were based on one man's body and marked on rules of wood or metal.

Each year the priests collected taxes from the farmers who paid in goods. To ensure that each farmer paid the same, there had to be standard jars to measure the grain, wine or oil and standard weights for weighing other produce. As the amount of tax each man paid depended on the size of his farm, the priests had to find a way to measure area. They did this by means of triangulation and this method has been used by surveyors ever since. These early Egyptians developed great practical knowledge of shapes, space and measurement.

East of Egypt was the great civilisation of Mesopotamia. Historians refer to the Sumerian, Chaldean, Assyrian and Babylonian stages of its development. In some ways the Mesopotamian civilisation was like that of Ancient Egypt, and the priests of both countries made great advances in astronomy. Mesopotamia had a large foreign trade and the travelling merchants needed standard weights and measures. To make trading easier, they used small bars of silver, each stamped with its weight. They had invented the world's first money.

In about 1500 B.C. trading centres developed on the eastern coast of the Mediterranean sea, in the land called Phoenicia. Phoenician seamen sailed all over the Mediterranean, trading goods. Later they made even longer voyages. During these voyages, the sailors studied the earth and sky and increased their knowledge of geometry (the study of earth measurement). They became great navigators and led the world in the science of navigation.

The early Greeks who lived on the coasts and islands of the Mediterranean also became sea traders and enjoyed travelling. This brought them wealth and knowledge from other lands. The Greeks enjoyed arguments and all new information was debated by rich citizens who had leisure time. Men skilled in argument had disciples who studied their methods. In this way, for example, Pythagoras taught his mathematical knowledge to a group of young men. Public study and debate led to new ways of thinking about mathematics, as every new rule was put to the test of debate. The science of geometry developed slowly into a vast logical system of rules. These were written into a series of textbooks by Euclid in about 300 B.C.

The word *arithmetic* is a Greek word, meaning "relating to numbers." The Greeks enjoyed solving number puzzles. But they had difficulty in the art of calculating as they used letters to stand for numbers.

The Romans conquered most of Europe but based their number system on that of the early Greeks. When the Roman Empire collapsed, the western world soon forgot the science of Greece and made no progress in calculation and science until knowledge came from the East.

One of the oldest civilisations in the world was in the valley of the Indus in India. The mathematicians of India were as knowledgeable as those of Mesopotamia and the merchants of India were as advanced in arithmetic. However, the Hindus developed a different sign for each number up to nine and made great progress in mathematics when they devised a dot to show an empty column. We now use a zero. This system enabled the Hindus to calculate with numerals. Mathematicians of Ancient India worked with fractions and wrote them down as we do. In about A.D.800, Indian traders took the knowledge of this new number system to Baghdad.

Baghdad was the Muslim capital of learning. Here the knowledge of Indian mathematicians and western scientific works (such as writings on astronomy and Euclid's geometry) were translated into Arabic by Muslim scholars.

In Muslim universities students studied Greek geometry, and Indian arithmetic as well as astronomy, trigonometry and geography. Gradually the Muslim mathematical knowledge spread to Europe. By A.D.1400 Arabic numerals were used in Europe and schools opened to teach the new arithmetic. Textbooks were written which showed a convenient way of writing down tables and problems. More important still, they introduced new arithmetical signs such as + (add) and − (subtract). Other signs, such as × (multiply), ÷ (divide), = (equals) were gradually introduced.

Long before the Arabic numerals reached Europe, Hindu and Muslim mathematicians had worked out rules for solving number problems. But although *algebra* is an Arabic word, Muslim mathematicians did not teach it as it is taught today. Algebra as we know it was not developed until about A.D.1600.

With the new mathematical knowledge and the beginnings of algebra, European mathematicians were equipped to tackle some of the practical problems of the modern world. As time passed, men built machines and workshops. Scientists studied the earth and sky, and a new means of calculation called *calculus* was invented. This is used every day by astronomers and engineers when they study motion and change.

As new problems are created by new developments, men invent new branches of mathematics to solve them. Progress in mathematics continues, but it is just as bound up with the problems of everyday life as it was thousands of years ago. Although today's problems are more complicated, modern mathematicians are fortunate to have electronic computers and other aids to help them make rapid calculations. Perhaps in the future there will be even greater discoveries. The story of mathematics is never-ending.

SI, Metric
and Imperial Tables

The International System of Metric Weights and Measures (SI)
Système International d'Unités

In recent years most countries have agreed internationally to adopt a common system of metric units. This is known as SI. It is based on earlier ideas and is, scientifically, an improvement on them. The six basic SI units are:

length	metre	(m)
mass	kilogramme	(kg)
time	second	(s)
electric current	ampere	(A)
temperature (absolute)	kelvin	(K)
luminous intensity	candela	(cd)

All the other SI units are derived from (or described in terms of) these six basic units, for example:

area	square metre	(m^2)
volume	cubic metre	(m^3)
velocity (speed)	metre per second	(ms^{-1})
force	newton	(N)
power	watt	(W)
electrical potential difference	volt	(V)

some SI Recommended Prefixes

x 1 000 000	mega	(M)	e.g. *megatonne*
x 1 000	kilo	(k)	e.g. *kilometre*
x 0.01	centi	(c)	e.g. *centimetre*
x 0.001	milli	(m)	e.g. *millilitre*

Table of Weights and Measures

Linear Measurement

Imperial Units

12 inches = 1 foot
3 feet = 1 yard
22 yards = 1 chain
10 chains = 1 furlong
8 furlongs = 1 mile

1 furlong = 22 yards
1 mile = 1,760 yards
1 mile = 5,280 feet

Metric Units

10 millimetres = 1 centimetre
10 centimetres = 1 decimetre
10 decimetres = 1 metre
10 metres = 1 decametre
10 decametres = 1 hectometre
10 hectometres = 1 kilometre

1 metre = 100 centimetres
1 kilometre = 1000 metres

Area

Imperial Units

144 square inches = 1 square foot
9 square feet = 1 square yard
4,840 square yards = 1 acre
640 acres = 1 square mile

Metric Units

100 square millimetres = 1 square centimetre
100 square centimetres = 1 square decimetre
100 square decimetres = 1 square metre
100 square metres = 1 are
10 ares = 1 dekare
10 dekares = 1 hectare
1 hectare = 10000 square metres

Volume

Imperial Units

1,728 cubic inches = 1 cubic foot
27 cubic feet = 1 cubic yard

Metric Units

1000 cubic centimetres = 1 cubic decimetre
1000 cubic decimetres = 1 cubic metre

Capacity

Imperial Units

5 fluid ounces	=	1 gill
4 gills	=	1 pint
2 pints	=	1 quart
4 quarts	=	1 gallon
2 gallons	=	1 peck
4 pecks	=	1 bushel
8 bushels	=	1 quarter
1 gallon	=	160 fluid ounces

Metric Units

10 millilitres	=	1 centilitre
10 centilitres	=	1 decilitre
10 decilitres	=	1 litre
100 litres	=	1 hectolitre
10 hectolitres	=	1 cubic metre
1000 millilitres	=	1 litre
1000 litres	=	1 cubic metre

Mass (Weight)

Imperial Units

16 drams	=	1 ounce
16 ounces	=	1 pound
14 pounds	=	1 stone
2 stones	=	1 quarter of a hundredweight
4 quarters	=	1 hundredweight
20 hundredweights	=	1 ton
1 hundredweight	=	112 pounds
1 ton	=	2,240 pounds

Metric Units

10 centigrams	=	1 decigram
10 decigrams	=	1 gram(me)
10 grams	=	1 decagram
10 decagrams	=	1 hectogram
10 hectograms	=	1 kilogram
100 kilograms	=	1 quintal
10 quintals	=	1 tonne (metric ton)
1000 grams	=	1 kilogram
1000 kilograms	=	1 tonne (metric ton)

Troy Weight

24 grains	=	1 pennyweight
20 pennyweights	=	1 ounce (troy)
12 ounces	=	1 pound (troy)
240 pennyweights	=	1 pound (troy)
5,760 grains	=	1 pound (troy)

Index

abacus 6
acre 6
add/addition 6
algebra 6
algebraic equation 7
altitude 7
angle 7, 64
angle of elevation 7
angle-meter 8
ante meridiem 8
apex 8
approximation 8
Arabic numerals 9
arbitrary measures 9
arc 9
area 9
arithmetic 10
astronomy 10
asymmetry 10
atomic clock 10
average 11
axis 11

balance 11
bar graph 11
base 12
bearing 12
billion 12
binary system 12
bisect 13
breadth 13
bushel 13

calculus 13
calendar 13
calliper 14
candela 14
capacity 14
carat 14
cardinal number 14
cardinal points 14
centigrade or
   Celsius scale 15
centilitre 15
centimetre 15
centre 15
chord 15
chronometer 15

circle 16
circumference 16
clinometer 16
closed plane figure 16
closed solid figure 16
common property 17
compass 17
composite number 17
concave 17
cone 18
conservation 18
continuous graph 18
convex 18
co-ordinates 19
count 19
cube 19
cubit 19
cuboid 19
cylinder 20

day 20
dead reckoning 20
decade 20
decagon 21
decilitre 21
decimal notation 21
decimetre 21
degree 21
denominator 22
diagonal 22
diagram 22
diameter 22
digit 22
digital machine 23
dimension 23
disc 23
division 23
divisor 24
dodecagon 24
dodecahedron 24

edge 24
elevation 7, 24
ellipse 25
empty set 25
enumerate 25
equation 7, 26
equiangular 26

equidistant 26
equilateral 26
equipoise 26
equivalent 27
estimate 27
even number 27

face 27
factor 27
Fahrenheit scale 28
fathom 28
figure 28
finite 28
foot 28
force 28
formula 29
fraction 29
furlong 29

gallon 29
gauge 30
geometry 30
gill 30
gradient 30
grain 31
gram (gramme) 31
graph 11, 18, 31
gravity 32
greater than 32
Greenwich meridian 32
gross 32

hand 33
hectare 33
height 33
heptagon 33
hexagon 34
hexahedron 34
histogram 34
horizontal 34
hour 35
hourglass 35
hundredweight 35
hyperbola 35
hypotenuse 36

icosahedron 36
imperial measures 36
inch 36
index 37
infinite 37

integer 37
interest 37
international system of
 units 38
intersection of sets 38
irrational number 38
isosceles triangle 38

Kelvin scale 38
kilogram 38
kilometre 39
kilowatt 39
knot 39

latitude 39
length 40
less than 40
line 40, 51
line segment 40
linear measure 40
litre 40
logarithm 40
longitude 41

"magic" square 41
magnetic pole 41
mapping 41
mass 42
matching 42
mathematics 42
mean-time 43
measure 9, 36, 40, 43
meridian 32, 43
metre 43
metric system 44
mile 44
milligram 44
millilitre 44
millimetre 44
minute 45
month 45
multiplication 45

natural number 46
navigation 46
net 46
nought 46
number 14, 17, 27, 38, 46,
 47, 48, 50, 57, 61, 62, 78
number line 47
number sentence 47

numeral 9, 47, 65
numerator 47

oblique 48
oblong 48
octagon 48
octahedron 48
odd number 49
odometer 49
one to one
 correspondence 49
open sentence 49
ordered pair 49
ordinal number 50
ordinality 50
ounce 50

palm 50
parabola 51
parallel lines 51
parallelogram 51
parentheses 52
partitioning 52
pack 52
pence 52
pennyweight 52
pentagon 53
pentagram 53
percentage 53
perch 53
perimeter 54
perpendicular 54
pi 54
pictogram 54
pie chart 55
pint 55
placeholder 55
place value 55
plane 56
plumb-line 56
polygon 56
polyhedron 56
post meridiem 57
pound 57
power 57
prime number 57
prism 58
probability 58
product 58
protractor 58
pyramid 59

Pythagorean theorem 59

quadrant 59
quadrilateral 60
quart 60
quire 60
quotient 60

radius 61
ratio 61
rational number 61
real numbers 61
rectangle 62
rectangle number 62
reflection 62
region 63
regular 63
relation 63
rhomboid 64
rhombus 64
right angle 64
right-angled triangle 64
rod 64
Roman numerals 65
rood 65

scale 65
score 65
second 65
section 66
sector 66
segment 40, 66
semicircle 67
sequence 67
set 25, 38, 67, 79
set square 68
sextant 68
SI 68
slide rule 68
solid shape 69
sphere 69
spiral 69
spirit level 70
square 41, 70
square of a number 70
square root 70
standard measures 71
statistics 71
stone 71
subset 72
subtraction 72

sundial 72
surface 73
surveying 73
symbols 73
symmetry 74

tally 74
tangram 74
tessellation 75
tetrahedron 75
theodolite 75
thermometer 75
three-dimensional 76
time 43, 76
ton 77
tonne 77
topological transformation 77
topology 77
triangle 38, 64, 78
triangle number 78
trigonometry 78
trillion 79
troy weight 79
truth set 79

two-dimensional 79

undecagon 80
union of two sets 80
unit 80
universal set 80

variable 81
Venn diagram 81
Vernier gauge 81
vertex 81
vertical 82
volume 82

watt 82
week 83
weigh 83
weight 79, 83
width 84

yard 84
year 84

zero 84